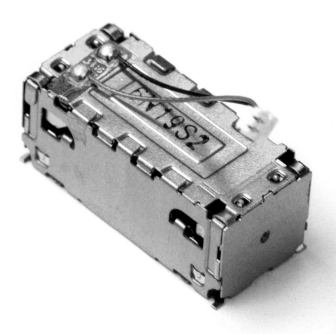

Teardown

Vol. 1

fictiv

fictiv

Fictiv is a manufacturing platform transforming how teams design, develop, and deliver the next generation of hardware products.

Content Lead
Christine Evans

Photographer
Meghan Baciu

Co-founder
Nate Evans

Co-founder: Dave Evans

Teardown Leads: Sylvia Wu, Sunny Sahota, Andrew Hudak, Richard Tran

Director of Community: Madelynn Martiniere

Contributing Designer: Constantin Nimigean

Designer: Ina Xi

Fictiv
1221 Mission St
San Francisco CA 94103

www.fictiv.com

Printed in Canada by Hemlock Printers Ltd.

ISBN: 978-0-692-87803-3

Contents

Foreword 10

Components of a Photograph 15

Apple Watch 17

Dyson Supersonic Hair Dryer 26

iRobot Roomba 35

GoPro Session 44

BB-8 Sphero 55

Fitbit Charge 2 63

Hoverboard 71

Lockitron Bolt 78

Ricoh Theta S 360° Camera 89

Nintendo Switch 98

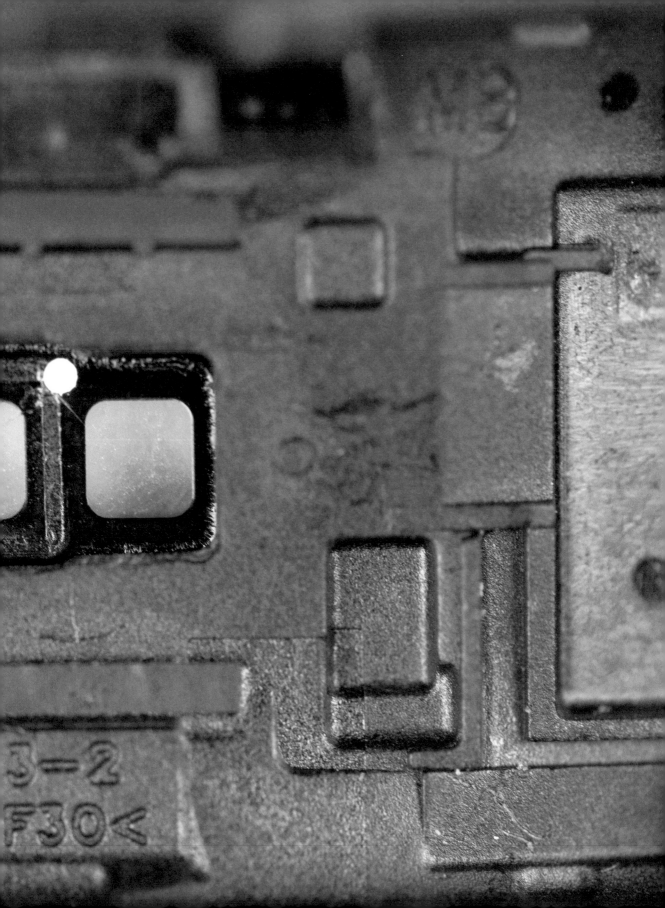

Foreword

By Christine Evans

Every physical product has a story. In its fully realized design, it may seem as if a product has always held its perfect form. In reality, a design is transformed, piece by piece, over a long and iterative journey.

When we did our first teardown at Fictiv almost two years ago, that seemed the most important part of it. By reducing a product to its smallest components, you see not only what it's made of (its materials, fasteners, and such) but also a glimpse into how it came to be.

You see the tool markings that reveal when the product left the factory. You see the carefully labeled installation marks that enabled it to flow through the assembly line. You see the ejector pin and gate marks left behind from injection molding. You see the design ingenuity behind a deceptively simple mechanism, or an expertly placed light pipe. You see an intricate marriage of human creativity and machine technology.

Fictiv certainly did not invent the concept of teardowns. Teardowns are common practice at hardware development companies as a way to uncover the design and engineering decisions that lie just beneath a product's glossy surface. In a dichotomous display of chaos and order, art and science, the individual vs the whole, each teardown reveals a new insight into how physical products come to be.

This collection was created to bring light to some of these key insights that give the most successful hardware companies a competitive edge. The photos chosen for this collection do not showcase comprehensive teardowns, but rather highlight some of the most compelling features for a range of well-known products.

My hope is that with this collection, you'll discover a new design perspective, a creative mechanical approach, or simply a spark of inspiration and creativity.

"There's a way to do better — find it."

— Thomas Edison

Composition:
rule of thirds

Symmetry

Depth of field

Balanced color palette
(warm vs cool tones)

Humanistic element

Contrasting elements:
sharp geometries vs
organic backdrop

Negative space

Focal plane

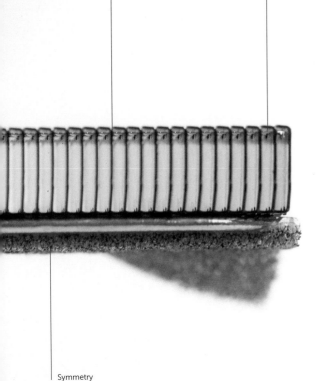

Symmetry

Components
of a Photograph

By Meghan Baciu

My favorite photography teacher once told me to approach every moment as a witness. A witness to the details of the environment without fixation on the resulting photograph — until right before I take it.

In photographing a teardown, my role is to capture both the technical detail as well as the beauty in each step. Every photograph is a point of discovery, to reveal the fascinating nuances of an object otherwise seen as ordinary.

And like the object itself, a successful photograph holds many different components in the balance: light, backdrop, composition, focus, and contrasting geometries, to name a few.

When these components align, they create an image that draws you in; makes you think; compels you to return.

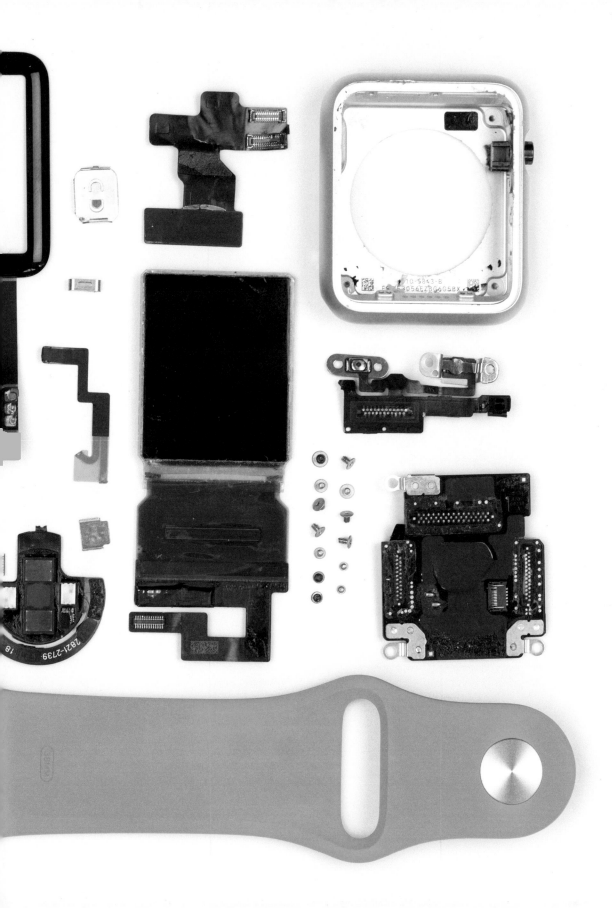

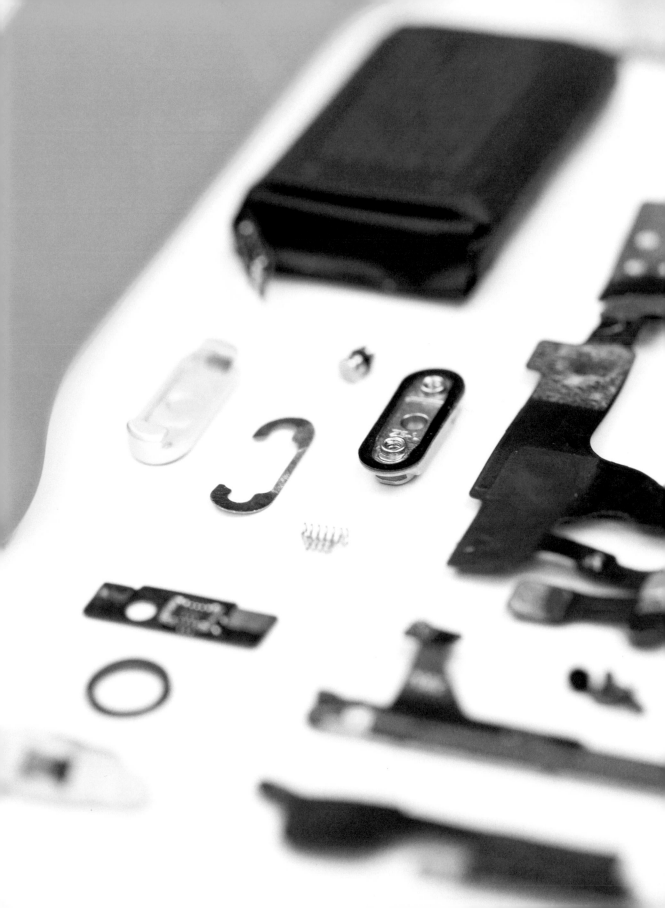

Apple Watch

Apple is well known for its meticulously designed products. From using custom aluminum alloy for the body, to a flexible OLED display, and intricate button and enclosure design, the Apple Watch displays strong attention to detail.

In the Apple Watch, there are two components we found most compelling:

Taptic Engine

The Taptic Engine is Apple's next generation component for haptic feedback, which is to use the sense of touch in a device interface to provide information to the user.

This custom component is an electromagnetic linear actuator that passes electric current through copper windings, generating magnetic force. This magnetic force moves permanent magnetic masses along a center rod.

Side Button Assembly

The watch's side button reveals an intricate multi-part assembly, developed to prevent water ingress at a moving interface.

When the button top is pressed, it nudges a small cylinder that in turn presses the dome button switch on the inside of the watch case. Meanwhile the mounting bracket remains stationary, and its o-ring forms a tight seal against the watch case.

Taptic engine

Side button switch

Digital crown

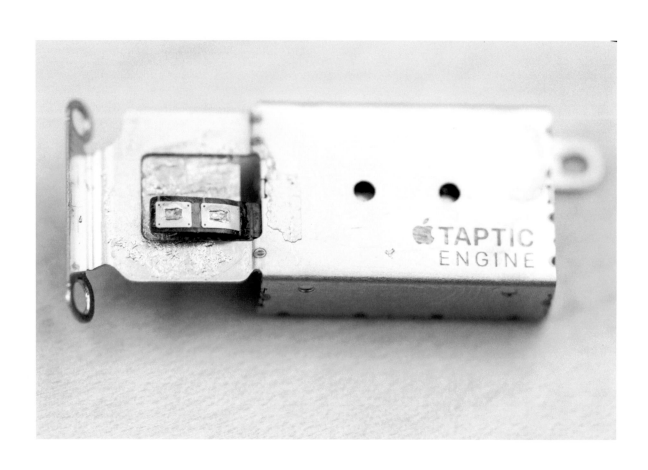

Center rod

Taptic engine

Copper windings

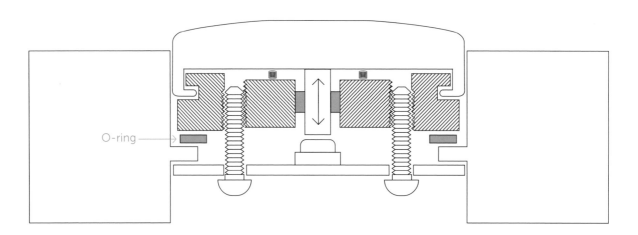

O-ring

Schematic of Apple Watch waterproof
button subassembly

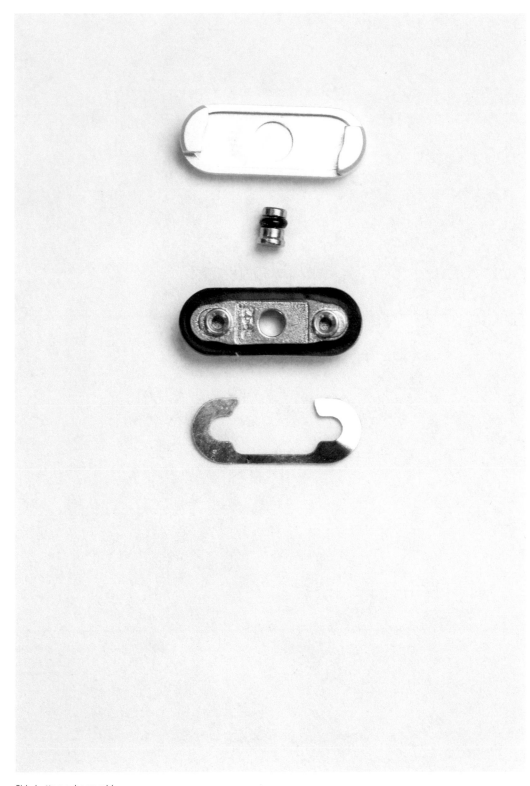

Side button subassembly,
designed for waterproofing

Dyson Supersonic Hair Dryer

We've long been fans of Dyson's work—their products are beautifully designed as well as mechanically impressive. Dyson is also a leader in transparency, one of Fictiv's core values, as they actively showcase how their products are made.

So we couldn't help but be intrigued when Dyson applied its technology to a new product category with the Supersonic Hair Dryer.

For this teardown, we were most intrigued by the supersonic motor, which is uniquely placed in the handle instead of the head. This placement is designed to improve the balance of the product, as well as reduce wrist movement.

Although not completely silent, the motor is significantly quieter than those found in conventional hair dryers. We analyzed the audio spectrogram to find no significant peaks in the audible range.

The brushless DC motor is encased in a silicone rubber sleeve, to reduce vibration against the handle, and the impeller attached to the motor shaft is machined on a five-axis mill.

Silicone rubber
sleeve

Impeller is machined on a 5-axis
mill out of single piece of aluminum

Dyson Supersonic Hair Dryer

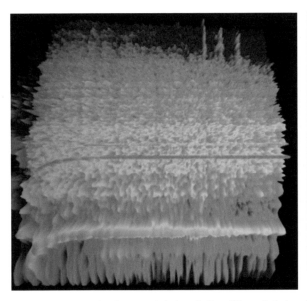 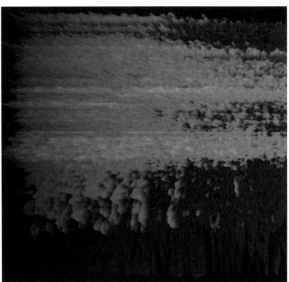

Audio spectrograms for drugstore hair dryer (left) and Dyson hair dryer (right)

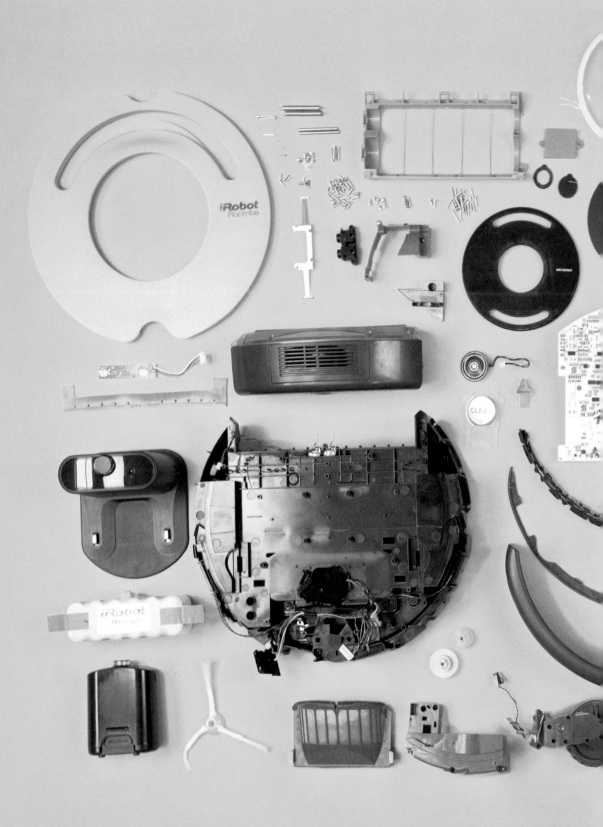

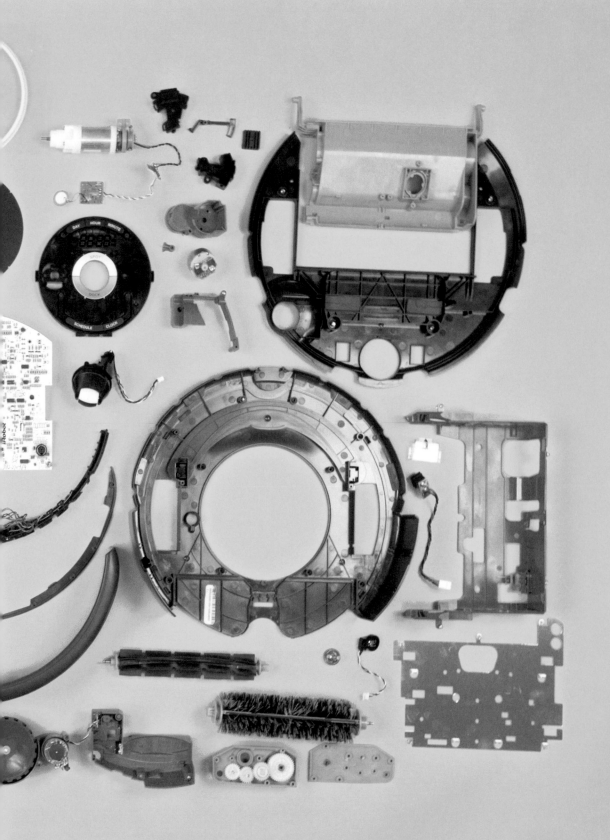

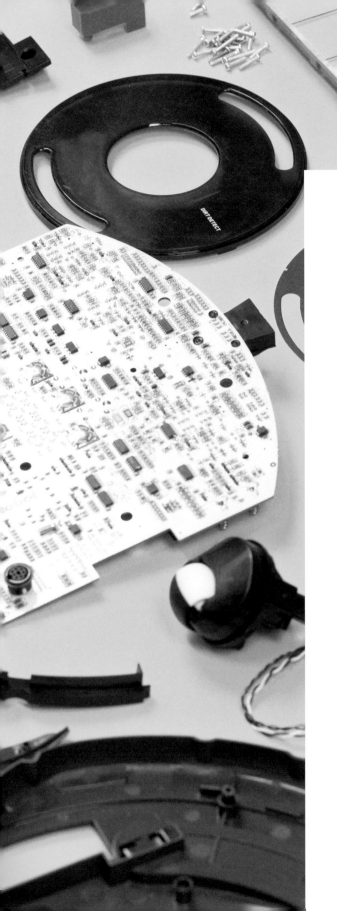

iRobot Roomba

The Jetsons predicted flying cars, jetpacks, and robot maids to serve the fortunate people of the 21st century. Only one of these predictions has achieved mass market adoption.

In addition to being a uniquely modular product for repairability, the Roomba holds an impressive array of navigational sensors. Here we highlight a few notable ones, including:

Caster Wheel Sensor

The caster wheel sensor is an optical encoder that tells the Roomba its direction of travel. Underneath the caster wheel is an IR (infrared) LED (light-emitting diode) and receiver pair.

Bumper Sensor

The Roomba knows when it's collided with an obstacle using its bumper sensors. The bumper sensor is quite big, but it's mostly empty space. A large, plastic, swinging arm assembly interrupts an LED-receiver pair in order to amplify any small impact to the bumper.

Top Sensor

The IR sensor on top of the Roomba helps the Roomba "see" far ahead, to identify the locations of iRobot's "Virtual Wall Lighthouse" accessories as well as the charging dock. We especially appreciate the intricate light pipe covering the IR sensor, which can detect light coming from all 360 degrees.

Caster wheel sensor

Bumper sensor

Top sensor + light pipe covering the IR sensor, which can detect light coming from all 360°.

IR receiver pairs on a
subassembly bracket

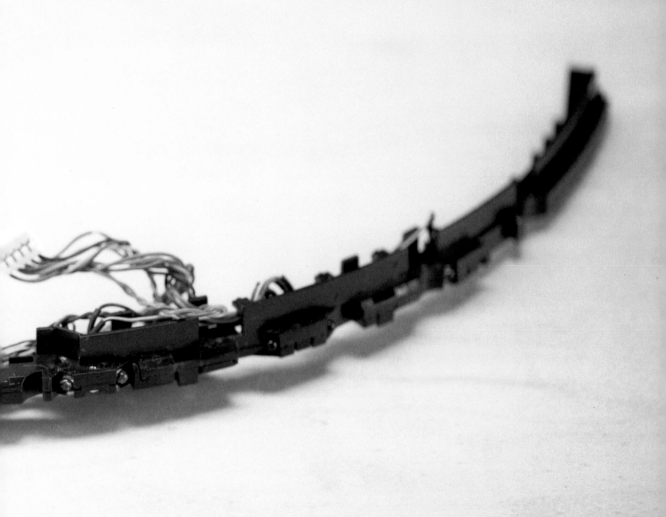

GoPro Session

One of the latest members of the GoPro family boasts compactness and robust waterproofing as major selling points. 50% smaller and 40% lighter than former generations and waterproof up to 30 feet, the Session is impressive.

The Session is incredibly compact, with a dense molded hard plastic shell providing a fully waterproof design without additional case accessories relied on by previous models.

With the enclosure removed (with difficulty), the Session reveals a neatly packed cube of boards and flex cables attached to the lens bracket. It appears that all of these components are fixed to the bracket with a generous amount of adhesive during assembly.

Like its predecessors, the f/2.8 lens is an asymmetrical glass build. On the back of the lens lies a heat spreader and thermal paste to reduce the load on the CMOS sensor, which is on the adjoining board.

Surprisingly, there's quite a bit of space between the lens and the surrounding components, despite other components being tightly packed within the cube—a rarity in the portable consumer electronics industry, where every tenth of a millimeter is sacred.

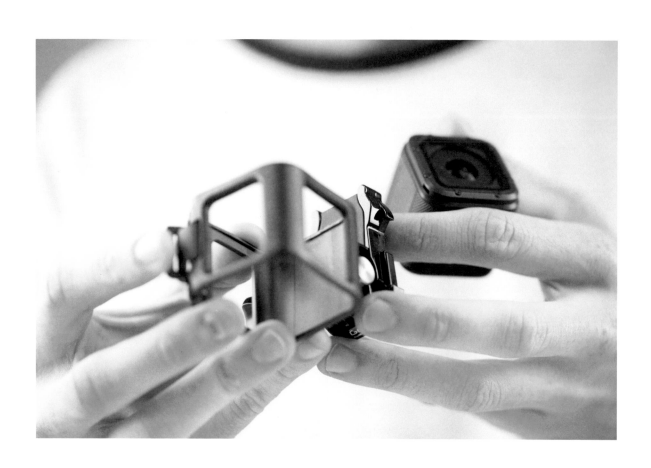

CMOS sensor, with thermal
paste to reduce the load

Adhesive acts as a waterproof
sealant over the buttons

GoPro Session

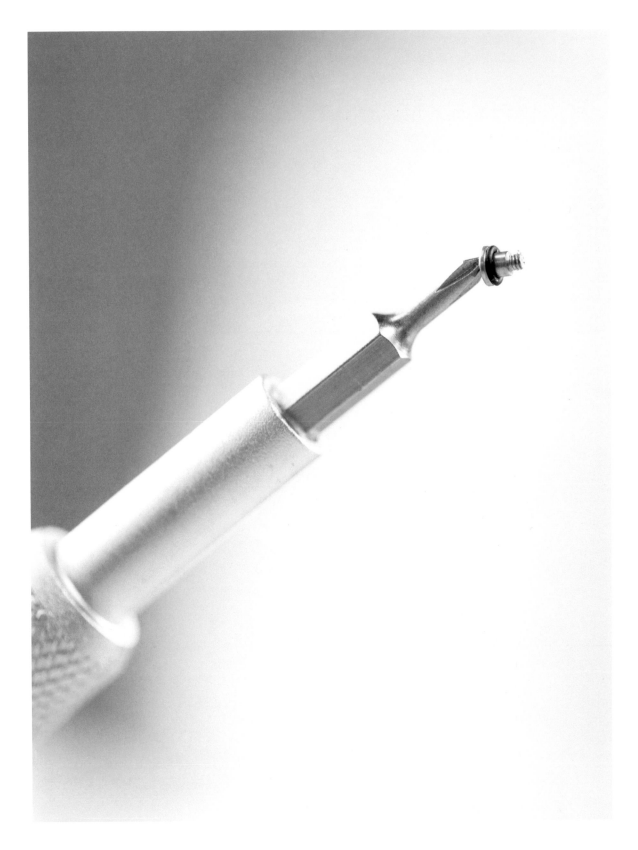

Mic

Light pipe

O-ring

BB-8 Sphero

BB-8 holds a special place in our hearts as the very first Fictiv teardown. This adorable droid was developed by Sphero in partnership with Disney and features Sphero's core technology, with the addition of a small BB-8 head held seamlessly on top with a magnet.

The major components of the BB-8 Sphero include:

Charging Dock

The charging dock is comprised of seven different plastic pieces that each interlock for a tight fit using snap fits and four bosses as a mounting point. The dock contains a PCBA as well as two weights, which are commonly found in consumer electronics products to make plastic parts feel more substantial.

Ball

The sphero ball is a 3mm polycarbonate shell and houses two standard motors used to drive each gear. The biggest change in this generation of the Sphero ball compared with previous generations is the spring-loaded mechanism with a magnet on top, designed to keep BB-8's head in place.

Head

The head is a simple polycarbonate shell with a magnet used to attach it to the ball. It's amazing how a simple addition like this really transformed the product without changing its core technology.

Spring-loaded magnet inside the Sphero
ball that holds up the BB-8 head

Weights inside the charging dock

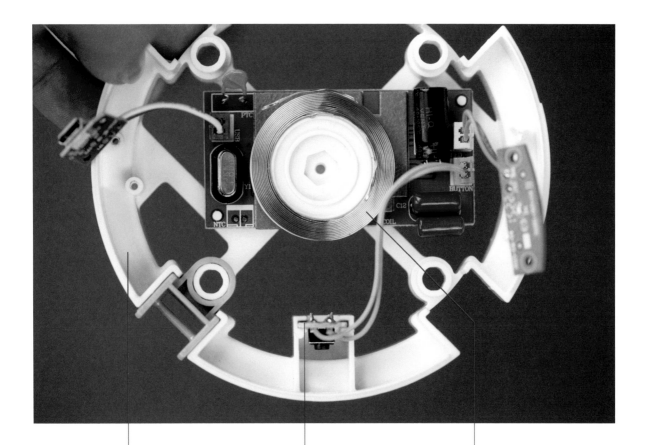

Single injection
molded part

Breakout board
with pushbutton

Inductive charging
coil

fitbit

63mAh

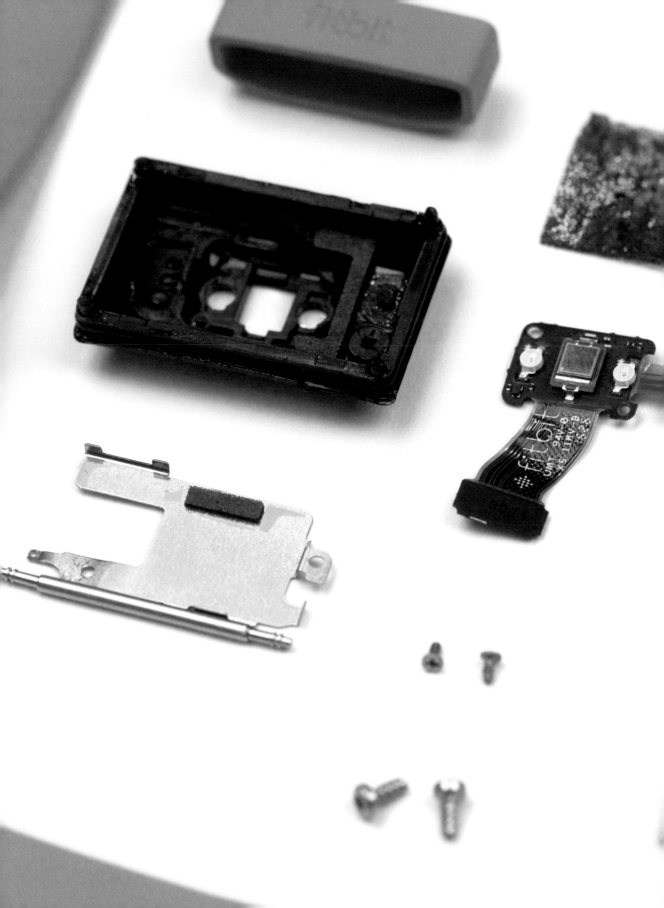

Fitbit Charge 2

For our Fitbit teardown, we wanted to understand more about three major features of interest for modern wearables:

Band to Body Attachment

The band of the Fitbit Charge 2 is unique in its use of a springy sheet metal stainless steel band attachment latch, which is fastened to a hard plastic insert on each band with a screw.

Space Management

The Charge 2 has a compact and efficient design. To reduce wasted space, the design limits the curve of the main body housing so that the components can be stacked, without much entanglement.

Waterproofing Techniques

The Charge 2 uses a combination of gaskets, o-rings, and adhesives to achieve a level of waterproofness in its product (making it sweat, rain, and "splash" proof). The screen bezel has a gasket which goes along the outside to fill the space between the bezel and the main body. This, in combination with adhesive covering the charging pins, achieves a watertight seal.

The button sub-assembly contains an o-ring that slides up and down as the button is depressed, to add an additional level of waterproofing. Lastly, an elastomer is used around the pressure sensor, which allows the sensor to be open to the environment for proper functionality while preventing water from entering the device.

Sheet metal band attachment latch

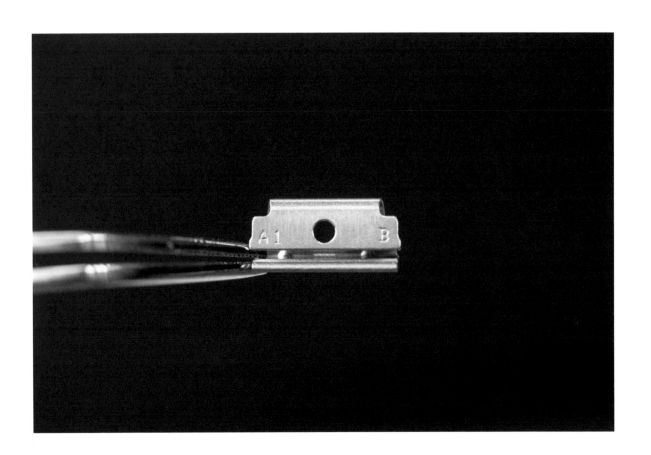

Compact button subassembly
for space management

Gasket used around
screen bezel

Waterproofing features

Button subassembly
with o-ring

Elastomer used around
the pressure sensor

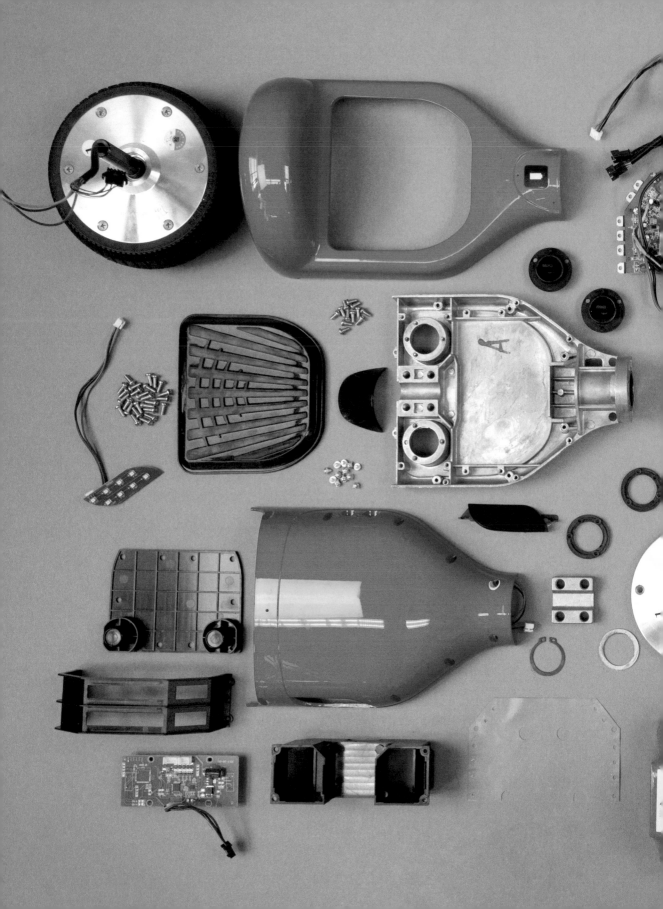

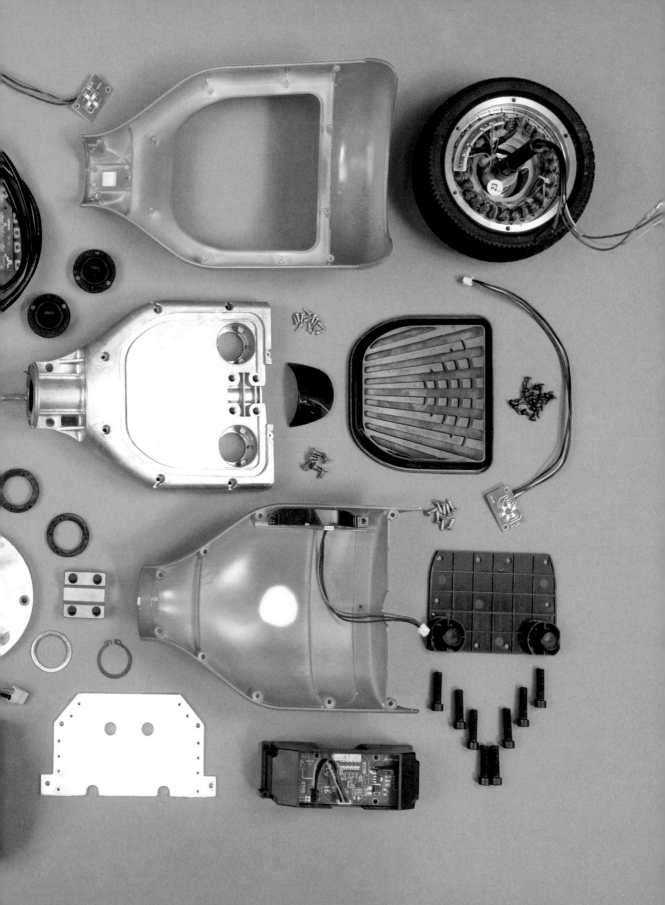

Hoverboard

The hoverboard was perhaps one of the biggest gadget phenomena of 2016. With its tumultuous history of fires, IP battles, and legislative restrictions, we were eager to see what made it so lucrative and polarizing alike.

What makes this product fascinating is that it combines several simple components that work together to create a highly unique product.

The motors are located within the wheels, making each wheel weigh about twelve pounds. The motor is a three-phase A/V, permanent-magnet brushless motor, which is often more efficient than D/C in high-powered devices.

While using spacers to stabilize the coils while keeping them compact against the core is a common practice in motor manufacturing, this manufacturer used a surprising workaround by using bamboo sticks (more commonly used as BBQ skewers) as the spacers. Certainly a unique off-the-shelf component!

What looks on the outside like a new, cutting-edge technology product is really just a couple of sensors and motors. The added space within the enclosures allows manufacturers to easily utilize off-the-shelf components, making it highly replicable and adaptable by the copycat market.

Main PCBA

Gyro PCBA

Battery

Cast aluminum frame

Bamboo stick used as a spacer between
the coil windings and their slots

Lockitron Bolt

Lockitron Bolt is a third generation smart lock product. For this teardown, we were thrilled to be joined by Lockitron founder Paul Gerhardt, where he shared key insights gleaned across the journey of bringing three product generations to market.

Paul was quite frank about how Lockitron overcomplicated their second-generation product with unnecessary features. Putting his learnings to practice, the design philosophy for this most recent product is rooted in ruthless simplification.

Off-the-Shelf vs. Custom

Here we highlight the balance between custom and off-the-shelf components to underscore Paul's main takeaway for new hardware companies:

The fewer custom parts you need, the less you have to pay for tooling. So for a new product with relatively low production volume, go with mostly off the shelf components and make as few custom parts as possible.

Worm gear

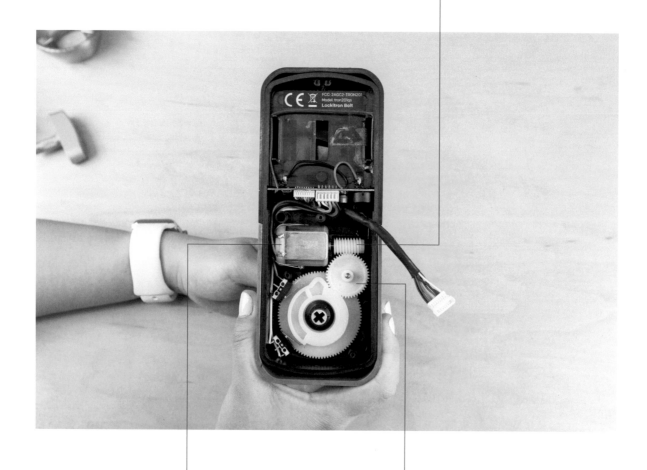

Reduction gear

DC motor

Standard lock components

Off-the-Shelf Components in Lockitron Bridge

The Electric Imp Wifi module
with BlinkUp

Bluegiga BLE112 bluetooth
low energy module

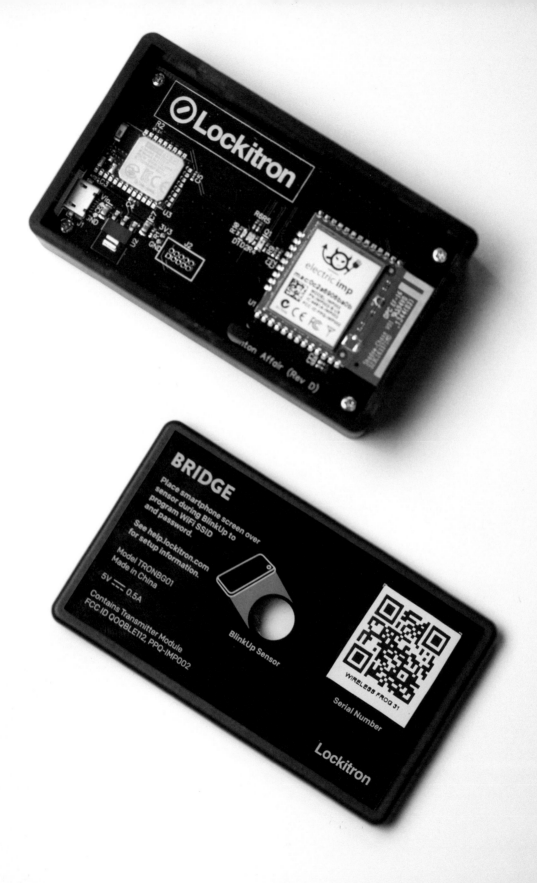

Custom injection molded delrin gears

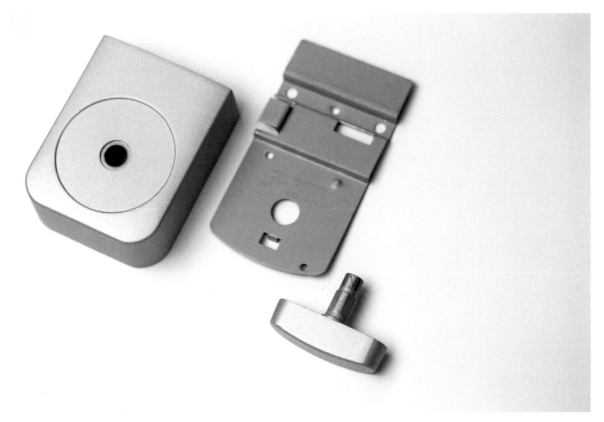

Custom cast aluminum housing, cast aluminum thumb turn, and
stamped steel mounting plate

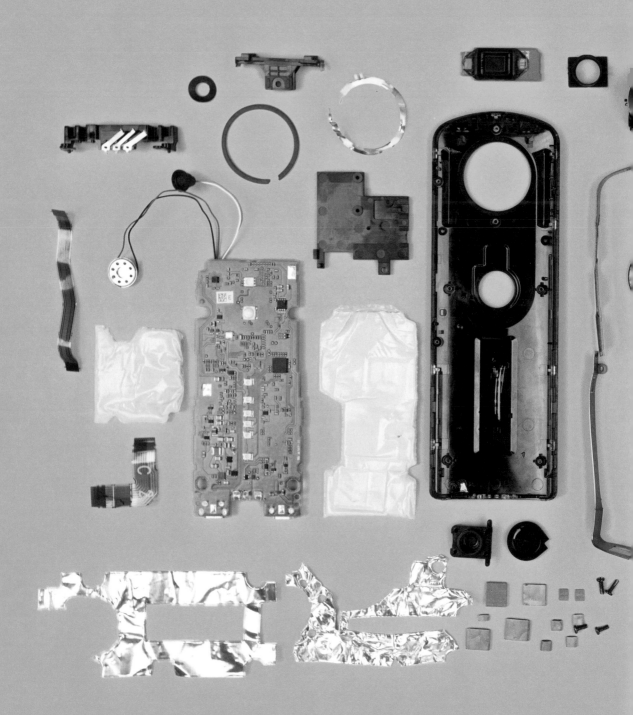

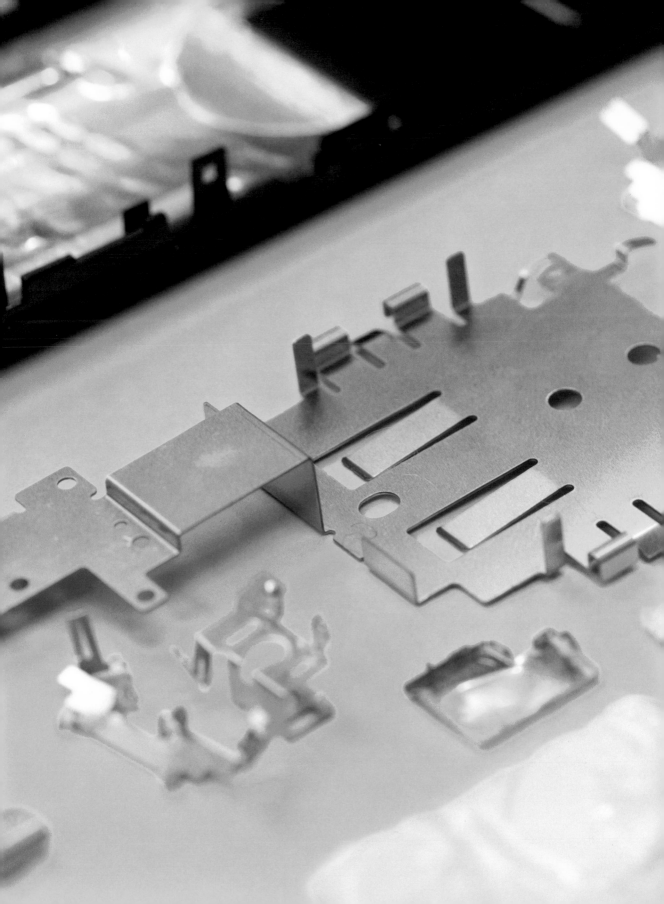

Ricoh Theta S
360° Camera

The Ricoh Theta S is an intricately designed 360° camera that captures panoramic images and video and then wirelessly sends the content to a mobile device.

The most unique and interesting part of this camera is no doubt the optics: two fisheye lenses are stacked back-to-back for 360° photo and video functionality. Images are "seen" by the fisheye lens using two prisms in the center of the optical assembly that "bend" the direction of light. Each side of the assembly features seven elements in six groups (if this sounds like a lot, it's actually not—the iPhone 6S has a five-element lens and a 3x optical zoom lens can have ten elements).

We were also impressed by the placement of the image sensors, which are geniusly positioned on the side of the optical train to minimize thickness.

The optical train is incredibly intricate and likely takes longer to assemble than some complete products. While our manufacturing mantra is to ruthlessly eliminate components, we can appreciate how much engineering and attention to detail went into this product.

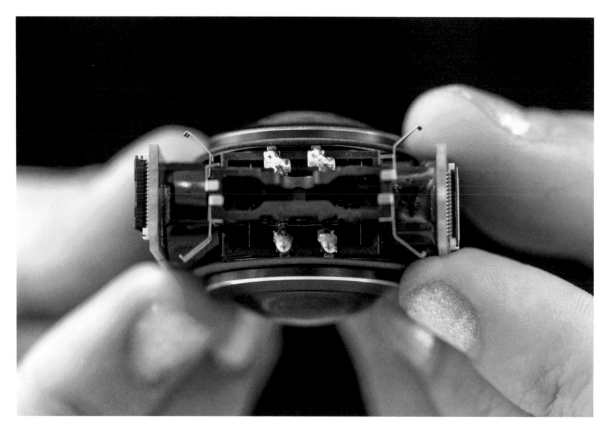

CMOS sensors placed on
the side of the optical train.

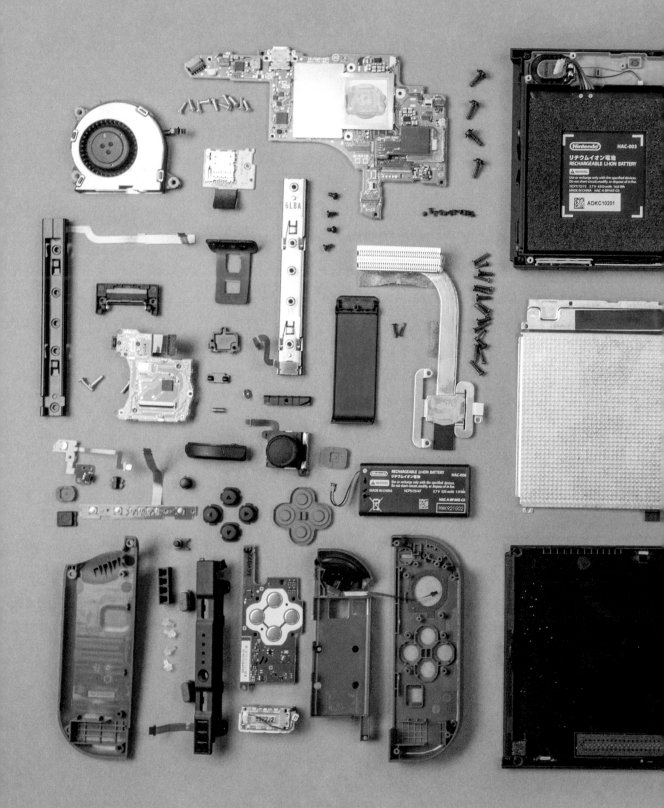

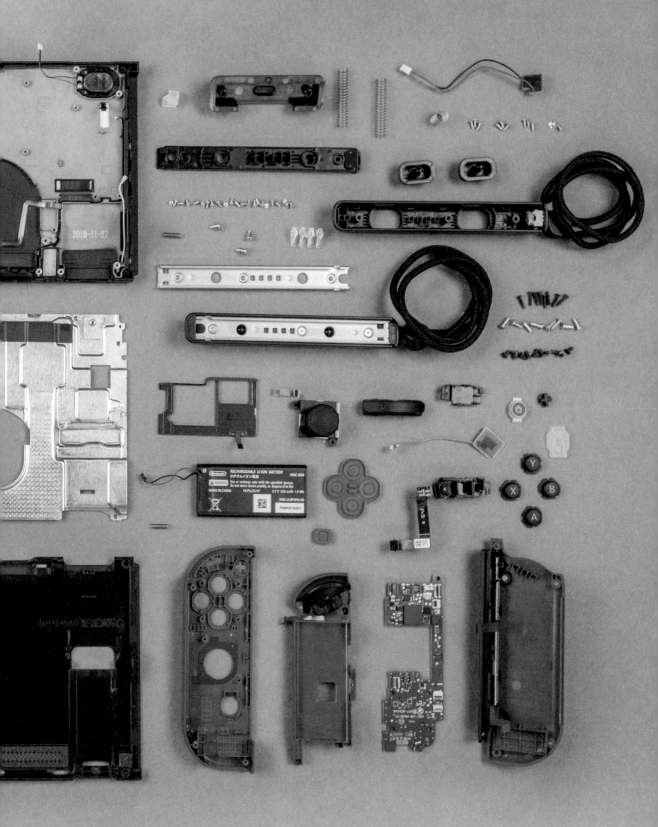

Nintendo Switch

The Nintendo Switch is a major evolution of the Wii platform, creating a hybrid home console and handheld mobile device. The platform's Lego-like modularity and expert component design showcase Nintendo's historical attention to detail.

HD Rumble Pack

Thanks to Apple leading the way with its Taptic Engine, everyone seems to be making investments in high-fidelity haptic feedback design. We found Nintendo's own version in each of the JoyCon controllers called the HD Rumble Pack, which is similarly a linear actuator.

Joystick

Compared to the joystick designs from previous Nintendo controllers, the Switch joystick is far smaller and lighter. With the lower profile design comes the added benefit of being more dust-proof.

Buttons

We were surprised to see that the conductive-pad style switches used in previous generations of Nintendo controllers were replaced with metal dome switches, with rigid button tops and an elastomer middle layer for spring back. The trigger and shoulder switches are now pushbutton switches.

Shoulder pushbutton uses two
compression springs for spring back

HD Rumble Pack

Bent sheet metal casing

Linear resonant actuator (LRA)

Internal mass travels along short axis

HD Rumble Pack

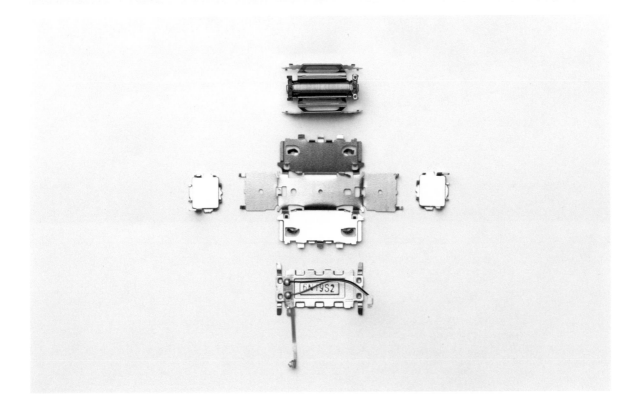

Nintendo switch joystick (left)
vs Gamecube joystick (right)

Nintendo Switch joystick
disassembled

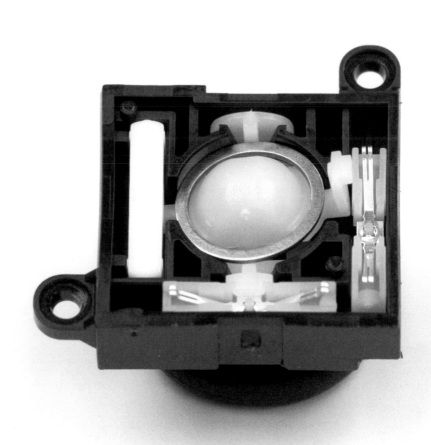

Nintendo Switch Joystick

Re-engineered to minimize height

FPC (flexible printed circuit) replaces a rigid PCB
as the joystick base

Sliding potentiometers in place of rotary potentiometers

Low profile metal dome switch instead of pushbutton

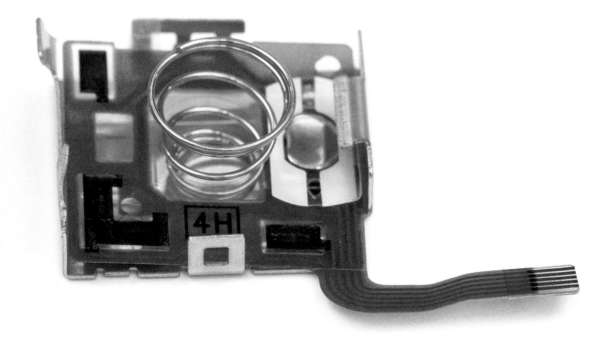

fictiv

Build better hardware.

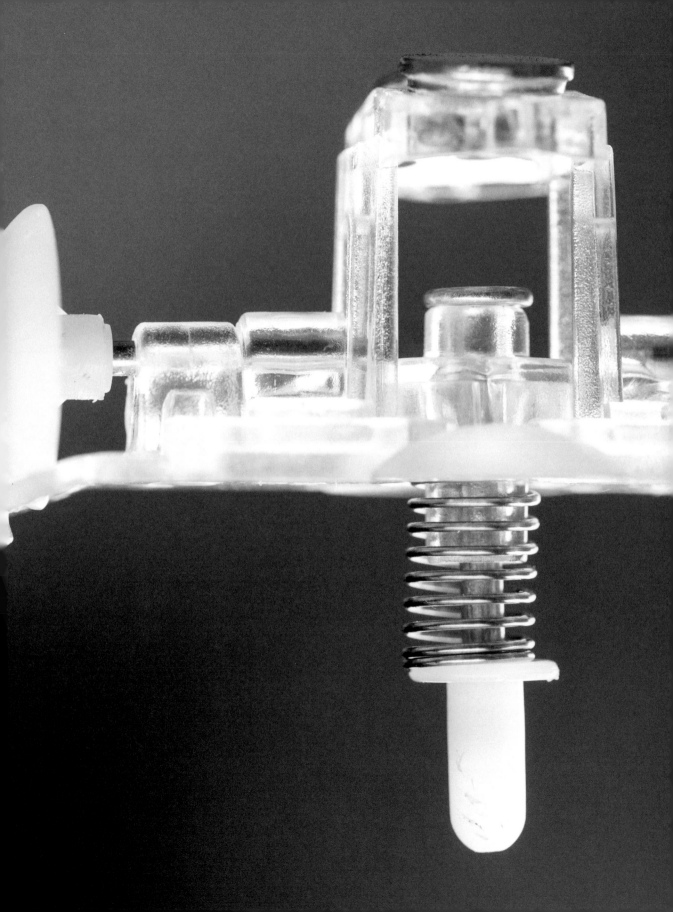